Hela MTIR

Breast cancer management for women over 70

Hela MTIR

Breast cancer management for women over 70

ScienciaScripts

Cover image: www.ingimage.com

This book is a translation from the original published under ISBN 978-620-6-73044-6.

Publisher:
Sciencia Scripts
is a trademark of
Dodo Books Indian Ocean Ltd. and OmniScriptum S.R.L publishing group

120 High Road, East Finchley, London, N2 9ED, United Kingdom
Str. Armeneasca 28/1, office 1, Chisinau MD-2012, Republic of Moldova, Europe
Managing Directors: Ieva Konstantinova, Victoria Ursu
info@omniscriptum.com

Printed at: see last page
ISBN: 978-620-8-64147-4

TABLE OF CONTENTS

INTRODUCTION

Breast cancer the most common cancer in women. Its incidence with age, occurring mainly between ages 50 and 60. Life expectancy steadily increasing, more so in women, and with it the number of cases of breast cancer, making breast cancer a public health problem. Breast cancer in elderly women presents a number of particularities, from epidemiological and clinical point of view, as well as in terms of treatment and prognosis. Although histologically more favourable, with less lymph node invasion, these tumours are diagnosed late. Patients aged over 70 are generally excluded from the majority of clinical trials, which explains the lack of specific recommendations for the management of this age group. The presence of varying degrees of co-morbidities in these patients leads to disparities in the provision of care, and gives life expectancy estimates considerable weight in therapeutic decisions. The actual age must be considered in relation to the physiological age and the presence or absence of defects, as there are frequent disparities between patients of the same chronological age. Unfortunately, this age group is, in the majority of cases, excluded from clinical research, both for screening purposes and for the study of new treatments. As a result, we cannot extrapolate from results obtained from studies that initially excluded them. In fact, the elderly population presents a number of particularities that can interfere with therapeutic management, starting with the estimated life expectancy, the social and psychological profiles specific to each individual, and the increased number of co-morbidities. As a result, the toxicity of chemotherapy products is increased by a number of mechanisms:

• Impaired glomerular filtration.

• Reduced liver volume and enzyme activity.

• A reduction in the distribution volumes of fat-soluble products due to a reduction in muscle mass and a relative increase in fat mass.

• An increase in the free fraction of drugs due to hypoalbuminemia.

We also note that age is often a factor in the downward revision of the care on offer, which seems unacceptable at present given the increase in female life expectancy and the development of new therapies that are adapted to and well tolerated by this population. Associated co-morbidities and respect for the wishes and autonomy of these patients are therefore the only factors to be taken into account when drawing up treatment plans. Despite the obvious link between age and breast cancer, the diagnosis is generally made later in the elderly (limited interest on the part of older women in self-examination or mammography, accessibility of screening centres, etc.). Secondly, breast cancer has biological characteristics that are specific to the elderly. For example, there are more tumours with hormone receptors (oestrogen and progesterone), and the expression of biological markers of aggressiveness (grade, percentage of S-phase cells, p-53 mutation and overexpression of cerb-B2) is less frequent.As for the histological type of tumour, there is a higher proportion of mucinous, lobular or papillary cancers than in younger women. Macroscopically, there are more tumours larger than 5 centimetres and less lymph node involvement. As a general rule, breast cancer in older women appears to be less aggressive, but this population is not immune to highly aggressive cancers or inflammatory cancers. Recent observations show that around 48% of patients aged 70 or over are diagnosed at an already metastatic stage.Once the diagnosis has been made, the main discussion centres on the choice of appropriate treatment. Between the ages of 35 and 55, breast cancer is the main cause of death in women, while from the age of 70 onwards, cardiovascular disease accounts for the greatest number of deaths. What's more, older people are often more reluctant to accept a more aggressive treatment regimen. When choosing which treatment to apply, it is therefore essential to take into account general factors such as physiological age, associated pathologies and the level of autonomy of the elderly person, so as not to 'over-treat' a person who may develop or worsen a severe pathology that will prove fatal. The aim of this study is identify the specific features of the management of breast cancer in patients aged over 70

MATERIALS AND METHODS

1. Literature review

The documentary research was carried out between 01/05/2023 and 15/05/2023.

1.1. Selection criteria

The criteria for selecting articles for the review are set out below.

Inclusion :

- Original research article or review.
- Dealing with the management of breast cancer in elderly subjects.
- Written in French or English.
- Published between 2018 and 2023

Non-inclusion criteria :

- Publication not corresponding to research work (e.g. teaching article) or method not described.
- Study focusing solely on a specific site (e.g. Medline) or system (e.g. Quick Clinical).

1.2. Search bibliography

We searched several bibliographic databases for published data.
PubMed/sciencedirect/ Wiley Online Library/ Springer/Google Scholar**:** We have identified relevantMeSH (Medical Subject Headings) Breast cancer
Treatment breast cancer eldery patients Cancer de sein sujet agées

We limited the search to articles written in English and French, published between 01/01/2018 and 15/05/2023.

1.3. Selection of articles

Based on the bibliographic records obtained by searching the databases, we carried out an initial sort according to title and abstract. The full-text versions of potentially eligible publications were retrieved, and an initial selection of articles for inclusion was drawn up. Secondly, we considered the bibliographic references of this first selection articles (excluding publications already selected). We retrieved the potentially eligible articles and made a second selection

There is no conflict of interest in our work.

DISCUSSION

1. Approaches to decision making in the treatment of breast cancer in elderly patients

Chronological age is not in itself a robust biomarker, and treatments standardised for younger patients should also be offered to older patients deemed fit enough to tolerate them. On the other hand, aggressive treatment can be avoided if the patient is more likely to die of other causes. We present the treatment decision tree in Figure 1 according to the National Comprehensive Cancer Network guideline reference [1]. Firstly, despite the difficulty of distinguishing "fit" from "vulnerable" and "frail" elderly patients, standard treatment should be offered to those considered fit. Secondly, the life expectancy of the elderly patient must be taken into account when determining appropriate treatments. |2], we need to ask whether breast cancer may or may not affect the patient's life expectancy. An online tool called "ePrognosis" is available, which is 10-year mortality prediction model that has been developed and validated in the USA |3]. In Japan, 48.0% of deaths in breast cancer patients aged over 75 were not cancer-specific (n = 27,385).

|4]. Life expectancy must therefore be taken account when choosing appropriate therapies, even if it is difficult to predict accurately. In the United States, the effect on life expectancy of a diagnosis of breast cancer in elderly women has also been studied using the Medicare Surveillance, Epidemiology and End Results (SEER) database. This was done by comparing the survival two groups, one consisting of 66,000 patients with breast cancer ≥67 years of age, and the other of

controls matched age, comorbidity, previous mammography use and social demographics [5]. This study revealed a counterintuitive result: women diagnosed with ductal carcinoma in situ (DCIS) or stage I invasive breast cancer actually had a lower risk of death than controls (hazard ratio; HR0.7, 95% confidence interval; CI 0.7-0.7 for DCIS HR 0.8, 95% CI 0.8-0.8 for stage I). In contrast, the risk of death was higher, regardless of age, for stage II patients (HR 1.1, 95% CI 1.0-1.2), and for stages III or IV, breast cancer itself was the most common cause of death [5]. Third, we consider whether or not the patient can tolerate standard treatment. Finally, we need to assess the patient's goals and values regarding cancer management and whether these compatible with cancer treatment. The management of any cancer presents particular difficulties in older people and in many cases treatment is refused because of age, although many can benefit if the right decision-making tools are used. As with any In the management of cancer in the elderly population, a comprehensive geriatric assessment (CGA) is of paramount importance.[68] Ideally, this assessment should include a geriatrician; however, if this is not possible, it can be carried out by an oncologist using widely available guidance and online assessment plans.[69] The CGE uses functional status around the patient's activities of daily living (ADLs & IADLS), gait assessment, visual/auditory status, performance statussocioeconomic status, psychological status, comorbidity scores, nutritional status, polypharmacy, and cognitive status.[70] The Vulnerable Elders SurveyVES-13 score in particular has been shown to predict functional decline and death in patients with early-stage breast cancer,[71] allowing oncologists to tailor management approaches and chemotherapy decisions based on baseline deficits.

Medication reconciliation, with attention to the indication of each drug, should be critically reviewed periodically to prevent adverse drug interaction events.

2. Aim of treatment

The aim of breast cancer treatment is to obtain :

• Locoregional control to avoid recurrence in the breast, wall or lymphatic territories.

• General disease control to avoid metastatic relapse.

• best functional and aesthetic results.

• side-effects from proposed treatments to improve quality of life

3. Therapeutic

3.1.Surgery

Breast cancer surgery is the mainstay of treatment. Its aims are to :

• Removal of the tumour, either by lumpectomy (conservative treatment) or by mastectomy (radical treatment);

• To enable a precise histological diagnosis;

• Biological analysis of the tumour, including hormone receptor research, S-phase measurement and cell proliferation marker assays;

• Analysis of the lymph nodes that drain the tumour (GS, or axillary lymph nodes). Removal of these lymph nodes improves local control of the disease and guides further treatment;

- Guide complementary treatments, which gives it a prognostic value.
- Minimise the aesthetic after-effects, sometimes by combining excision with plastic surgery to leave a normal-looking breast in the case of conservative treatment, or to perform breast reconstruction (immediate or delayed) in the case of radical treatment.

Mortality from breast cancer surgery in elderly women is very low, ranging from 0 to 0.3% [6].

Even in old age, standard surgery should be offered, given the benefits in terms of morbidity and mortality and the low rate of post-surgical complications [7, 8].

Advanced chronological age was considered to be a poor prognostic and predictive factor in surgical oncology [72,73].

In the mid-1990s, on the basis functional age, co-morbidities, difference in physiology, response to anaesthesia and stress, older patients were refused appropriate surgical treatment more often than younger patients [74].

3.1.1. Conservative surgery

Conservative treatment is defined as complete removal of the tumour surrounded by margins of healthy glandular tissue of at least 3 mm. Depending on the authors and the extent of the margins, it is referred to as lumpectomy, enlarged lumpectomy, quadrantectomy or partial mastectomy [89]. The extent of the lumpectomy must be adapted to the size of the tumour and may be supplemented by recuts. Quadrantectomy involves wider surgery with a minimum macroscopic margin of 2 cm, which most often corresponds to the removal of one quadrant. There is no real consensus on the margins of excision to be achieved during lumpectomy, but histology must confirm that healthy tissue has been removed. However, we try to minimise the aesthetic after-effects. Thanks to local radiotherapy, which reduces the risk of recurrence, conservative treatment has become the gold standard for early breast cancer. Older patients often choose mastectomy rather than breast-conserving surgery in Japan [9], unlike in other countries [10, 11] In Morocco, women over 70 prefer conservative surgery to a radical procedure [11]. For them, the breast remains a symbol of femininity and many elderly women remain very "attached" to their breasts, In the United States, most elderly women opt for mastectomy for fear of recurrence [75].

Lumpectomy is currently the standard treatment for invasive tumours:

- less than 30 mm long, with a tumour volume/breast volume ratio that allows sufficient excision and a satisfactory aesthetic result;
- single vision ;
- located at a distance from the areola ;
- without inflammatory signs;
- without an extensive intracanal component. [12]

3.1.2. Radical surgery

Classically known as the Patey mastectomy, this is a modified radical mastectomy described by Madden in 1972, with monobloc removal of the breast, including the skin sheath and areola-nipple plate, and axillary curage of the first two Berg stages, with preservation of the 2 pectoral muscles (unlike the historical Halsted operation, which sacrificed the 2 pectoral muscles, or the Patey operation, which preserved the pectoralis major but resected the pectoralis minor) [13].

The rules governing the use of surgery are, in the majority of cases, the same as those for younger women. As a result, mastectomy is indicated for large tumours, in cases of multi-focality and in cases of recurrence after conservative treatment. Although older patients with comorbidities have a higher risk of postoperative complications, these do not in themselves lead to higher relative mortality [14]. Frail patients who refuse surgery or have life-limiting comorbidities are unlikely to be candidates for surgery. According to a study presented at the 12th European Conference on Breast Cancer, healthy women over 70 can benefit from surgery, while their less fit and more frail counterparts can be treated with oral hormone therapy alone[79,80]. Although data is limited. Surgery for DCIS (ductal carcinoma in situ) may into account life expectancy and histological grade [15]. Healthy patients with high-grade DCIS should undergo surgery [16]; for low- and intermediate-grade DCIS, omission of surgery may be considered. In older patients with operable oestrogen receptor-positive breast cancer, surgery [alone or followed by tamoxifen (TAM)] was compared with TAM alone in the Cochrane database [17,18], and there was no significant difference in overall survival, but a significant improvement in progression-free survival. Subsequently, surgery has been recommended in healthy patients with low morbidity rates for the management of local control. In fragile patients or those with locally advanced tumours, primary endocrine therapy is recommended to allow less invasive surgery. In elderly patients hormone receptor-negative breast cancer who are not candidates for surgery or who refuse it, observation may be recommended rather than chemotherapy, particularly in asymptomatic cases, until local symptoms appear or cancer-related problems are expected, although there are few data to support this recommendation. In these cases, alternative palliative local therapy, non-surgical removal or irradiation, could be offered rather than conservation alone for local control. Reconstruction could be offered if desired by the patient, although the advantages and disadvantages of complex versus simpler procedures need to be assessed and discussed with patients [19].

3.1.3. Lymph node surgery

• Axillary surgery :

Lymph node investigation is an integral part of breast cancer surgery. Its main purpose is to enable histological analysis of the lymph nodes, in order establish the prognosis and subsequent therapeutic strategy. It also helps to reduce the risk of axillary recurrence by removing metastatic lymph nodes.

Lymph node dissection is systematic for all invasive tumours. It is standard practice to perform axillary lymph node dissection with at least 10 lymph nodes, including Berg's level I (subpectoral) and II (retropectoral). [13]

Various authors have noted that the indication for lymph node curage decreases with advancing age, compared with younger patients [20]. In fact, Pappo found that only 15.1% of lymph node curage procedures were carried out in patients over 70, compared with 30.3% in younger patients [21]. A number of reasons have been put forward for this, including an increase in post-operative morbidity and the higher prescription hormone therapy in elderly women, given the high rate of positive hormone receptors in this age group. Given the morbidity of axillary lymph node dissection (lymphoedema, shoulder stiffness, sensory disorders, etc.), it seemed necessary to find an alternative that would provide the same information about possible metastatic lymph node invasion, but with fewer consequences. This led to the development of the sentinel lymph node concept.

• Sentinel node :

The sentinel lymph node (SLN) technique for breast cancer was first described in 1993 by Krag and Weaver. The sentinel node technique makes it possible offer conservative treatment of axillary lymph nodes in patients with small breast cancer. size. The principle of the technique is to remove only one lymph node, known as the "sentinel" lymph node, which reliably reflects the lymph node status of the axilla. To locate this node, it must first be marked with a lymphophilic product (dye and/or isotope). Three detection methods have been proposed: colorimetric, isotopic or combined. The main objective is to reduce upper limb morbidity associated with axillary lymph node dissection [22]. If it is negative, lymph node dissection is not performed, whereas lymph node invasion indicates the need for additional lymph node dissection.

This technique is indicated, with the patient's informed consent, in the following cases:

- no clinical axillary involvement: N0 ;
- small T1 tumour (< 2cm), whether palpable or not
- unifocal tumour ;
- invasive cancer diagnosed prior to surgery
- no neoadjuvant chemotherapy or initial irradiation ;
- breast and armpit that have not undergone prior surgery. [22]

The prognostic value of the sentinel lymph node is comparable in older and younger women [23].

the Society of Surgical Oncology recommends that sentinel node biopsy should not be routinely performed in patients aged 70 or over with hormone receptor-positive, clinically node-negative breast cancer, based on evidence that axillary surgery in this population does not affect outcome [24] in frail patients with limited survival, neither the sentinel gg technique nor axillary dissection may always be necessary, particularly in those with clinically lymph node-negative, hormone receptor-positive breast cancer. Sentinel lymph node biopsy in clinically node-negative women over 70 with hormone receptor-positive, HER-2-negative early breast cancer is not recommended by the Choosing Wisely campaign and the Society of Surgical Oncology, as it is unlikely to change management [76,77].On the other hand, in healthy elderly patients, the sentinel gg technique may be acceptable because it is a minimally invasive surgical procedure and life expectancy is expected to be longer.

Martelli et al [83] showed in their publication that there was no benefit to axillary surgery in elderly patients with early-stage breast cancer whose nodes are clinically negative, in terms of breast cancer mortality. They went further and stated that even sentinel node biopsy could be abandoned. Their final conclusion was that axillary curage should be limited to the small number of patients who later develop overt axillary disease. In addition, Boughey [84] et al. stated in their publication that the time may have come stop surgical staging of the axilla altogether in hormone receptor-positive women over the age of 70.

3.1.4. Oncoplastic surgery and breast reconstruction

The combination of plastic surgery and cancer surgery has enabled considerable progress to be made in the treatment of breast cancer, hence the new terminology of oncoplastic surgery. The various techniques have improved aesthetic results and are tending to broaden the indications for conservative treatment. The contribution of these techniques means that in the case of mastectomy, immediate or deferred reconstruction can be systematically proposed; and in the case of conservative treatment :

- wide resections
- reduce the rate of local recurrences
- improve aesthetic results
- extend the indications for conservative treatment to lesions previously treated by mastectomy.

As for breast amputations, it is now permissible to offer breast reconstruction. This does not alter the patient's carcinological surveillance, does not increase the risk of local or metastatic recurrence, and allows mastectomised women to regain their bodily integrity and reduce the psychological impact of such mutilation. Initially performed at a distance from the mastectomy (secondary breast reconstruction), it is increasingly performed immediately, at the same time as the operation (immediate breast reconstruction), and is particularly indicated for women who will not undergo post-operative radiotherapy. Mastery of all reconstruction techniques is an essential prerequisite for being able to offer each patient the individual solution best suited to her clinical presentation and her needs [25].

Reconstruction can be offered if the patient wishes, although the advantages and disadvantages of complex versus simpler procedures need to be assessed and discussed with patients. [26].

3.1.5. Complications

Complications of locoregional breast surgery are essentially related to lymph node dissection. These complications are not more frequent in older women than in younger women, as confirmed by Djordjevic [27] and August [28] who found 35% of complications in younger patients compared with 25% in older patients. These complications were minor. They were mainly wound superinfections and lymphoceles.

• Peroperative :

o Axillary vein wound

o Transection of the serratus major or dorsalis major nerve

o Section of the inferior scapular pedicle, contraindicating reconstruction using a dorsalis major flap

• Immediate postoperative :

-Lymphorrhoea and lymphocele: these are common, and only constitute a complication when they are prolonged. No preventive measure is certain. It may require punctures after removal of the drains.

-Parietal infections or lymphorrhoea.

-Sensory disturbances of the thorax and inner arm.

• Late postoperative :

They account for most of the morbidity associated with axillectomy.

-Lymphoedema: its frequency is estimated at 2 to 20% [29]. Modern, more conservative techniques tend to reduce its frequency. It is more frequent in cases combined radiosurgery on the axillary fossa.
-Lymphangitis triggered by an injury to the upper limb must be treated without delay.
-Stiffness of the shoulder that can be avoided with good physiotherapy.

There is evidence that women aged 80 and over tolerate breast cancer surgery well overall, with low complication rates depending on their comorbidities. Delayed healing the only complication noted in 6% of cases in a study of 129 women 80 and over. Thirty-two per cent underwent simple mastectomy, 27% breast conservation and 6% axillary dissection [78].

3.2. Radiotherapy

Radiotherapy plays a major role in optimising locoregional control, in particularsince the advent of conservative treatment. It is most often combined with surgery as a post-operative treatment or, exceptionally, may be used exclusively. It is consistently effective regardless age, but more so in the case of significant factors in locoregional recurrence, in particular axillary lymph node invasion. However, irradiation of the elderly must be considered from two complementary angles, one concerning the absolute indications for radiotherapy, the other dealing with the immediate or delayed tolerance of irradiation. Unlike surgery, there are no absolute contraindications to radiotherapy. However, mental disorders and profound changes in general condition may preclude treatment. There are two main constraints in elderly patients: Radiotherapy requires a minimum level of cooperation from the patient in preparing for the treatment and during the treatment . Maintaining an immobile position, usually supine, is essential to the quality of the treatment, which requires good reproducibility of radiation parameters from one session to the next. This means that significant impairment of higher functions or major neurological disorders inevitably affect the quality of the treatment.

• Conventional irradiation (8 to 10 Gy per week over 4 to 5 weeks) often causes both physical and psychological asthenia in elderly patients, which is all the more marked the further they live from the treatment centre.

For this reason, hypofractionated radiotherapy regimens can be used to facilitate irradiation of the elderly, particularly in the case of comorbidities, as the elderly woman will not have time to develop complications. The Courdi study evaluated hypofractionated radiotherapy regimens and hormone therapy in 115 women with a median age of 83 for whom surgery was not indicated due to refusal, locally advanced tumour or co-morbidities. The 5-year recurrence-free survival was found to be 78%, leading the authors to consider hormone therapy and hypofractionated radiotherapy as alternatives to surgery for patients who refused surgery or conventional irradiation regimens [30]. Partial irradiation techniques are based on the fact that most local recurrences occur close to the initial tumour, but are still being evaluated. Locoregional tolerance to irradiation is no different to that of younger women. Precautions regarding dermatological care are identical, with particular attention paid to folded areas and the mammary fold in patients whose breasts often pendulate [31]. All in all, any elderly person in good general condition should benefit from a additional radiotherapy depending on the usual indications, as inadequate treatment has an adverse long-term effect on survival. Radiotherapy (RT) is generally well tolerated and is also recommended for older women [32]. It should be noted that the risk local recurrence is lower in older women and that the benefits of RT after conservative treatment diminish age [33]. In the randomised controlled trial for elderly patients with hormone receptor-positive early breast cancer, RT after conservative surgery may be omitted [34]. In Japan, 41.3% of more than 12,000 patients aged 75 or over who underwent conservative treatment received RT [35]. On the other hand, in older patients with high-risk breast cancer, breast RT may have a significant beneficial effect on relapse-free survival [36].

In a randomised trial (CALGB 9343) of women aged 70 and over with T1N0M0 tumours, the rate of locoregional recurrence after breast conservation was 10% in those who received tamoxifen alone, compared with 2% in those who received tamoxifen plus radiotherapy [81]. Overall survival at ten years was almost identical, with most patients dying of causes other than breast cancer.A systematic review and meta-analysis by Chesney et al [85] demonstrated the superiority of the combination of radiotherapy and tamoxifen over tamoxifen alone. They concluded that in patients over 70, radiotherapy reduces the risk of breast and axillary recurrence, but has no impact on distant recurrence or overall survival early breast cancer treated with breast-conserving surgery and tamoxifen. Intraoperative radiotherapy has also been studied and Vaidya et al [86] recently published a larger randomised controlled trial comparing targeted intraoperative radiotherapy (TARGIT) with conventional postoperative radiotherapy after breast conserving surgery in women with early breast cancer. They reported that in breast cancer patients over 45 years of age with hormone-sensitive invasive ductal carcinoma less than or equal to 3.5 cm in size, TARGIT, combined with lumpectomy as part of a risk-adapted approach, is as effective, safer and less costly than postoperative radiotherapy. Life expectancy is therefore an important factor in determining the real benefit of RT after conservative surgery, as are the risk factors for breast cancer.Post-mastectomy radiotherapy with one to three positive nodes remains controversial and decision making could then be guided by the relevance age, life expectancy, comorbidities, tumour burden and biology [37] All in all, any elderly patient in satisfactory general condition should be able to benefit from additional radiotherapy, depending on the general indications [38]. Inadequate treatment has an adverse long-term effect on survival [39, 40]. The benefit of radiotherapy is greater for women under 50, but remains significant even for those over 70 (risk reduction from 13% to 3%) [41].

Primary RT without surgery or chemotherapy in unfit elderly patients may be an option, with some retrospective data suggesting durable local control at doses above 60 Gy. [82].

3.3.Systemic adjuvant therapy for elderly patients

3.3.1. HER2-negative breast cancer (luminal type and triple negative type)

Luminal breast cancer (RH+/HER2-) accounts for 70.0% of breast cancers patients over 75 and 64.1% in those aged 55 to 64 [35]. Triple-negative breast cancer (RH-/HER2-) accounts for 14.4% of all breast cancers in patients aged over 75 and 13.6% in those aged 55-64 [35].In the Japanese study [16] of stage II/III patients, only 9.2% of patients over 75 received chemotherapy for luminal disease, while 32.9% received chemotherapy for triple-negative disease [42].

In randomised trials of older patients, standard regimens of cyclophosphamide, methotrexate and fluorouracil (CMF) or doxorubicin plus cyclophosphamide (AC) were superior to capecitabine and were therefore recommended for older patients with HER2-negative breast cancer [43.44]. Similar to other randomised clinical trials, docetaxel plus cyclophosphamide (TC) was superior to AC in both older and younger patients with tolerable adverse event profiles, although older patients experienced more febrile neutropenia with TC and more anaemia with AC [45. 46]Retrospective data from the National Cancer Database (NCDB) in the US indicate that the effect of adding chemotherapy to local treatment on the sentinel node technique in elderly women with triple-negative breast cancer is significantly beneficial chemotherapy administration after adjusting for weight, comorbidity and tumour factors using propensity matching analysis |47] This could make it possible to consider chemotherapy in the treatment of patients aged 70 or over with high-risk luminal breast cancer or triple-negative breast cancer.

3.3.2. Recurrence score in elderly breast cancer

To identify patients likely to benefit from chemotherapy in luminal cancer, the 21-gene recurrence score (RS) has been used to make more informed treatment decisions [48]. The RS test has been validated by including elderly patients, although even in one large validation study, the TAILORx trial for node-negative cancers, only 7% of patients were over 70 in the low-risk cohort and 4% in the intermediate-risk cohort [49] and the RxPONDER trial for node-positive cancers, only 12% of patients were over 70 [50]. In the US SEER database, among 147,107 patients aged 70 and over (elderly cohort) with HR-positive breast cancer, 11,476 patients () underwent RS testing, compared with 67,191 (18%) in patients aged 18-69 (young cohort) (P<0.001) [51]. In the older cohort, 1824 patients (16%) had high-risk RS, of whom 489 patients (52%) received chemotherapy, compared with 73% in the younger cohort [51].Although if the patient is too frail to receive chemotherapy, RS testing may be unnecessary because it would not change clinical management, the benefit of chemotherapy would help determine the efficacy of treatment modalities and could help guide shared decision making in healthy or vulnerable elderly patients.

3.3.3. HER2-positive breast cancer

In Japan, the incidence of HER2-enriched breast cancer (RH-/HER2+) 5.4% in patients over 75, 9.8% in those aged 55-64 and 5.3% in patients aged over 75 and 8.in those aged 55-64 for luminal-HER2 breast cancer (RH+/HER2+) [9]. Anti-HER2 therapies are the standard of care in elderly patients, for example trastuzumab [52, 53], pertuzumab [54] and trastuzumab emtansine (T-DM1) [55], although specific data for elderly patients are scarce due to their under-representation in clinical trials. In addition to chemotherapies, it is still important to demonstrate that they are beneficial for elderly patients, without the toxicity caused by chemotherapy. To this end, a randomised clinical trial was conducted to investigate trastuzumab monotherapy versus a combination of trastuzumab and chemotherapy in women aged 70 or older with HER2-positive primary breast cancer in Japan (RESPECT study, NCT01104935Fig. 2) [56] The initial hypothesis was that trastuzumab monotherapy would not be significantly inferior to trastuzumab combined with chemotherapy in terms of survival. (DFS) and at the same time superior in terms of safety and quality of life. However, the primary objective of non-inferiority of monotherapy was not achieved in terms of HR, although the reduction in chemotherapy-free survival compared with combination therapy was small. It was therefore concluded that, given the lower toxicity and good HRQoL profile, trastuzumab monotherapy may be a viable option for adjuvant treatment in selected elderly women with breast cancer [56] T-DM1 is an antibody-drug conjugate that combines the HER2-targeting properties of trastuzumab with cytotoxic activity of the microtubule inhibitory agent DM1. When tested in the pivotal KATHERINE trial of adjuvant T-DM1 in patients with HER2-positive early breast cancer and residual invasive disease after completion of neoadjuvant therapy, a significantly reduced risk of disease-free survival was reported compared with trastuzumab alone [57]. It should be noted that of the 743 patients enrolled in the trial, only 126 were over 65 years of age and the HR was 0.55 (95% CI: 0.22-1.34).

3.4.Chemotherapy

Older women with breast cancer are under-represented in clinical trials, and data on the effects of adjuvant chemotherapy in these patients are scarce. Oncologists are often concerned that older women may not be able to tolerate more intensive chemotherapy and therefore may not always receive care according to treatment guidelines; such shortcomings can have a negative impact on survival [88].Although adjuvant chemotherapy has improved survival in women with early breast cancer [89], the Oxford Overview 15-year analysis included too few patients over the age of 70 to accurately assess the effect of chemotherapy in this age . Healthy older women with breast cancer can tolerate chemotherapy as well as younger patients [90] and the more severe toxicities of chemotherapy in older patients did not significantly affect the benefits of adjuvant chemotherapy [91]. Various chemotherapy toxicity assessment tools are available to provide an estimate of the expected toxicity of chemotherapy regimens in elderly patients, including the Chemotherapy Risk Assessment Scale for Elderly Patients (CRASH)[92] and the CARG chemotherapy toxicity calculators.[93]

Cardiac function (anthracyclines, trastuzumab), neuropathy (taxanes), renal function and myelodysplasia (anthracyclines) are important co-morbidities to consider before choosing chemotherapy options.

For older women in good health who are candidates for adjuvant chemotherapy, the combination of CMF (Cytoxan, methotrexate, 5-fluorouracil [5FU]) remains the most widely used regimen but is generally not well tolerated. Replacing 5FU with an oral agent, capecitabine, has not been shown to be effective in preventing relapse [94]. The standard first-line regimen of doxorubicin plus cyclophosphamide (AC) is generally not an option for elderly patients due to anthracycline-related cardiotoxicity. Taxanes, such as nab-paclitaxel or docetaxel plus cytophosphamide (TC), are generally well tolerated, and 4 cycles of adjuvant TC regimen are better tolerated than AC regimen in elderly and young patients, with more febrile neutropenia but less anaemia [95]. Haematological toxicities leading to dose delays, interruptions and hospitalisations remain a problem in elderly patients receiving cytotoxic chemotherapy. The use of the 21-gene recurrence score (Oncotype dx score) to guide chemotherapy decisions is well established in patients with early-stage node-negative HR-positive breast cancer. However, this score is not well validated in elderly patients. The use of chemotherapy was not associated with better survival in older patients than in younger patients with a high-risk Oncotype dx score in a study of the SEER (Surveillance, Epidemiology, and End Results) database [96], which included a small proportion of older patients. Some experts recommend the addition of chemotherapy for patients with high-risk characteristics and a life expectancy of more than 5 years [97].HER-2-targeted agents have shown to improve survival in patients with HER-2-amplified breast cancer; however, all major trials were under-represented in patients over 65 years of age. The NSABP-31 trial, which compared AC/T trastuzumab, had approximately 16% of patients over the of 60, with significant improvement in disease-free survival in all age groups [98]. A systematic review a 47% relative risk reduction with use of trastuzumab in patients over 60 years of age, but a 5% rate of clinically significant cardiotoxicity in patients receiving trastuzumab [99].70 years who received trastuzumab, but were associated with concomitant use of anthracyclines [100]. Overall, there is good evidence that adjuvant or neoadjuvant chemotherapy prolongs survival in older breast cancer patients. The UK-based PREDICT model and Adjuvant! Online provide 5- and 10-year survival estimates of benefit of adding chemotherapy, endocrine therapy or HER-2-targeted therapy based on a range of demographic and clinical factors [101,102].

3.5. Systemic treatment for elderly patients with metastatic breast cancer

Although the principles of management of elderly patients are similar to those of younger patients, the differences are short life expectancy, some co-morbidity, drug interactions and relatively poor functional status. The aims of treatment for metastatic breast cancer are to control the cancer for as long as possible and to maintain the functional level and quality of life at work as as in younger patients [58]. A full work-up is useful to assess the condition and establish an individualised treatment plan, as outlined in the section below.

In patients with metastatic HR-positive breast cancer, the first choice is endocrine therapy, which has less toxicity and a good quality of life. combination of a cyclin-dependent kinase 4/6 inhibitor is also effective and probably safe in older patients [59]. Chemotherapy is also being considered for elderly patients with life-threatening or rapidly progressing visceral disease. When using chemotherapy in elderly patients, it is necessary to monitor toxicity and adjust doses [60].

In patients with metastatic triple-negative breast cancer, chemotherapy may be considered for those with rapidly progressing symptoms or visceral disease. Sequential single-agent cytotoxic chemotherapy may be offered with careful dose adjustment for toxicity, as there is concern about impending organ dysfunction in the elderly patient. In patients whose tumour expresses programmed cell death ligand 1 (PD-L1), the addition of an immune checkpoint inhibitor to chemotherapy has been recommended over chemotherapy alone, although evidence-based recommendations are limited in older patients. Pembrolizumab is approved in combination with chemotherapy (nab- paclitaxel,

paclitaxel orgemcitabine-plus-carboplatin), for patients whose tumours express PD-L1 with a combined positive score ≥ 10[61]

Atezolizumabis has also been approved in Japan in combination with nab-paclitaxel for patients whose tumours express PD-L1 with a combined positive score of≥ 1% [62]. In these pivotal trials, recruitment of older patients was limited. Although age-associated immune system decline or immunosenescence may alter the efficacy of immune therapy, pooled analysis of elderly Japanese patients with PD-L1-positive advanced non-small cell lung cancer indicated that efficacy and safety outcomes with pembrolizumab were similar to those in the general population [63]. Secondly, an immune checkpoint inhibitor could be offered to older patients with PD-L1-positive triple-negative breast cancer, if the patient can receive standard chemotherapy, although real-world data are needed in this population.

In patients with HER2-positive metastatic breast cancer, chemotherapy combined with anti-HER2 therapies is the standard treatment, regardless of age [64]. , for older patients, the benefit of first-line trastuzumab-based regimens was demonstrated in the RegistHER study, which was a large observational study including women aged 65 and over compared with women who did not receive trastuzumab, treatment plus trastuzumab was associated with a significant improvement in PFS, although OS was not significantly different [65]. In elderly patients, the issue is whether or not they can receive standard chemotherapy with anti-HER2 targeted therapies. In elderly patients, it may be difficult to maintain sufficient intensity of the standard dose of chemotherapy due serious adverse events, particularly standard chemotherapy. On the other hand, T-DM1 may be feasible and effective in elderly patients, as discussed above [54] Secondly, in Japan, a randomised controlled trial has been conducted and is currently recruiting to study trastuzumab, pertuzumab and docetaxel versus T-DM1 in patients aged 65 years or older with HER2-positive advanced breast cancer [Japan Clinical Oncology Group (JCOG)1607 HERB TEA study; UMIN000030783, Fig. Prior to the start of this study, a survey of Japanese physicians was conducted regarding the initial dose of docetaxel, with most recommending 60 mg/m2 for women aged 75 years or older [66]. The primary endpoint was OS to confirm the non-inferiority of T-DM1 with fewer toxicities. Regarding another less toxic regimen, a prospective phase II study confirmed the benefit adding metronomic chemotherapy (oral cyclophosphamide 50 mg daily) to pertuzumab plus trastuzumab in elderly women with an acceptable safety profile (EORTC 75111- 10114) [67] Therefore, elderly patients who are unfit for standard chemotherapy and frail may still benefit from anti-HER2 targeted therapies with a less intensive chemotherapy regimen.

4. Care considerations support

Bone health is of paramount importance in the elderly patient population and is even more important given the risks associated with breast cancer bone metastases and treatment-related bone loss. Osteoclast inhibition significantly reduces the risk skeletal-related events in women bone metastases; the choice of agent (i.e. a bisphosphonate or denosumab) depends on many factors. The balance between the risk of preventing osteoporosis-related fractures through the use of bisphosphonates and the additional risk of atypical fractures needs to be carefully considered [87].

There are also many other areas of support that an oncogeriatric cancer patient may need to maintain the best possible quality of life. These areas are best identified using the CGA discussed above and should be addressed early in treatment planning and management. falls through durable medical equipment and home safety assessments, managing nausea and vomiting, frequent consideration of polypharmacy and reducing the number of tablets are just some of the most important useful measures. Early involvement of palliative care may be essential for older patients with advanced/metastatic breast cancer who have a poor track record of health service use.

In summary, elderly breast cancer patients represent a unique challenge and an opportunity to provide the best possible care in line with their goals. As Mahatma Gandhi and Arti Hurria, who championed geriatric patients, said, "the true measure of a society is how it treats its most vulnerable members". Elderly breast cancer patients need maximum care, empathy, love and competent support to overcome this miserable disease during the autumn years of their lives.

CONCLUSION

There is no doubt that life expectancy increases, older patients need a new approach to the treatment of their breast cancer. Under-treatment of elderly breast cancer patients is risky when it is based solely chronological age, and hormonal manipulation alone can lead treatment resistance. Therefore, every elderly breast cancer should be properly assessed and all available treatment options offered to patients accordingly, unless there are contraindications. Case-by-case treatment in this age group would be more appropriate depending on the patient's assessment, co-morbidities and wishes. In addition, elderly patients should be taken into account in clinical trials, as they have been very poorly represented to date. These trials are important for analysing treatment and response in this age group and for improving treatment modalities for these patients. On the other hand, we need to ensure that we do not expose frail elderly patients to the side effects of treatment, particularly chemotherapy, if the risks outweigh the benefits in terms of survival. Multidisciplinary meetings should now incorporate tools such as ePrognosis and the Charlson Index into decision-making to maximise the survival benefits for these patients. In the meantime, a paradigm shift is needed in elderly breast cancer patients to minimise under-treatment and maximise survival.

REFERENCES

1-Older Adult Oncology Version: 1.2021. NCCN Clinical Practice Guidelines in Oncology (NCCN guidelines®) https://www.nccn.org/guideli nes/guidelines detail?category=4&id=1452 (4 January 2022, date last accessed)

2Extermann M, Brain E, Canin B, et al. Priorities for the global advancement of care for older adults with cancer: an update of the International Society of Geriatric Oncology Priorities Initiative. Lancet Oncol 2021;22:e29-36

3ePrognosis; Estimating Prognosis for Elders. https://eprognosisucsfedu/ (24 January 2022, date last accessed)

4Sawaki M, Yamada A, Kumamaru H, et al. Clinicopathological characteristics, practical treatments, prognosis, and clinical issues of older breast cancer patients in Japan. Breast Cancer 2021;28:1-

8Sawaki M, Yamada A, Kumamaru H, et al. Clinicopathological characteristics, practical treatments, prognosis, and clinical issues of older breast cancer patients in Japan. Breast Cancer 2021;28:1-8

5Schonberg MA, Marcantonio ER, Ngo L, Li D, Silliman RA, McCarthy EP. Causes of death and relative survival of older women after a breast cancer diagnosis. J Clin Oncol 2011;29:1570-7

6HURRIA A, COHEN A-J Practical geriatric oncology. Cambridge University press; 2010;p3- 109

7Audisio RA, Pope D, Ramesh HS, et al. Shall we operate? Preoperative assessment in elderly cancer patients (PACE) can help. A SIOG surgical task force prospective study. Crit Rev Oncol Hematol 2008;65:156-63.

8. Crivellari D, Aapro M, Leonard R, et al. Breast cancer in the elderly. J Clin Oncol 2007;25:1882-90

9Sawaki M, Yamada A, Kumamaru H, et al. Clinicopathological characteristics, practical treatments, prognosis, and clinical issues of older breast cancer patients in Japan. Breast Cancer 2021;28:1-8

10Crivellari D, Aapro M, Leonard R, et al. Breast cancer in the elderly. J Clin Oncol 2007;25:1882-90.

11. Sandison AJ, Gold DM, Wright P, Jones PA. Breast conservation or mastectomy: treatment choice of women aged 70 years and older. Br J Surg 1996;83:994-6.

12COTHIER-SAVEY I, RIMAREIX F, BELICHARD C. General principles of oncoplastic surgery and immediate and delayed breast reconstruction. Encycl. Med. Chir (Elsevier, Paris), Gynaecology; 871-A-25, 2002, 14p

13CLOUGH K-B, HEITZ D, SALMON R-J. Loco-regional surgery for breast cancer Encycl. Med. Chir (Elsevier, Paris), Gynaecology ;41-970, 2003, 15p

14e Glas NA, Kiderlen M, Bastiaannet E, et al. Postoperative complications and survival of elderly breast cancer patients: a FOCUS study analysis. Breast Cancer Res Treat 2013;138:561- 9

15Ward EP, Weiss A, Blair SL. Incidence and treatments of DCIS in octogenarians: grade matters. Breast Cancer Res Treat 2017;165:403-9

16japanese Journal of Clinical Oncology, 2022, 52(7)682-689 https://doi.org/10.1093/jjco/hyac054 Advance Access Publication Date: 23 April 2022 Review Articl

17Hind D, Wyld L, Beverley CB, Reed MW. Surgery versus primary endocrine therapy for operable primary breast cancer in elderly women (70 years plus). Cochrane Database Syst Rev 2006;1:CD004272.
18. Hind D, Wyld L, Reed MW. Surgery, with or without tamoxifen, vs tamoxifen alone for older women with operable breast cancer: cochrane review. Br J Cancer 2007;96:1025-9
19 James R, McCulley SJ, Macmillan RD. Oncoplastic and reconstructive breast surgery in the elderly. Br J Surg 2015;102:480-8
20 TRUONG P-T, BERNSTEIN V, WAI E, CHUA B, SPEERS C, OLIVOTTO I-A.

Age-related variations in the use of axillary dissection: a survival analysi of 8038 women withT1-ST2 breast cancer Int J radiation oncology Biol. Phys 2002;54:794-803
21 PAPPO I, KARNI T, SANDBANK J, DINUR I, SELLA A, STAHL-KENT V, et al.

Breast cancer in the elderly: histological, hormonal and surgical characteristics The breast 2007;16:60-7
22 BENAMOR M, NOS C, FRENEAUX P, CLOUGH K. Sentinel node technique in breast cancer Encycl. Med. Chir (Elsevier, Paris), Gynécologie ;865-F-10, 2004, 15p 23 GENNARI R, ROTMENSZ N, PEREGO E. Sentinel node biopsy in elderly breast cancer patients Surg Oncol. 2004;13:193-196.
24 Society of Surgical Oncology. https://www.choosingwisely.org/clinician-li sts/sso-sentinel- node-biopsy-in-node-negative-women-70-and-over/ (12 December 2022, date last accessed) 25 COTHIER-SAVEY I, RIMAREIX F,

BELICHARD C. General principles of oncoplastic surgery and immediate and delayed breast reconstruction. Encycl. Med. Chir (Elsevier, Paris), Gynaecology; 871-A-25, 2002, 14p.
26 ames R, McCulley SJ, Macmillan RD. Oncoplastic and reconstructive breast surgery in the elderly. Br J Surg 2015;102:480-8.
27 DJORDJEVIC N, KARANIKOLIC A, PESIC M. Breast cancer in elderly women Archives of gerontology and geriatrics 2004 ;39:291-9
28 AUGUST D-A, REA T, SONDAK V-K Age-related differences in breast cancer treatment Am. Surg. Oncol. 1994 ;2 :180-8
29CLOUGH K-B, HEITZ D, SALMON R-J. Loco-regional surgery for breast cancer Encycl. Med. Chir (Elsevier, Paris), Gynaecology ;41-970, 2003, 15p
30COURDI A, ORTHOLAN C, HANNOUN-LEVI J-M, FERRERO J-M, LARGILLIER R, BALUMAESTRO C, et al. Long-term results of hypofractionated radiotherapy and hormonal therapy without surgery for breast cancer in elderly patients. Radiotherapy and oncology 2006;79 :156-61.

31SERIN D, ESCOUTE M. Breast cancer in elderly women Encycl Med chir, Gynécologie; 689-A-20, 1999, 6p
32Kunkler IH, Audisio R, Belkacemi Y, et al. Review of current best practice and priorities for research in radiation oncology for elderly patients with cancer: the International Society of Geriatric Oncology (SIOG) task force. Ann Oncol 2014;25:2134-4
33Smith BD, Gross CP, Smith GL, Galusha DH, Bekelman JE, Haffty BG. Effectiveness of radiation therapy for older women with early breast cancer. J Natl Cancer Inst 2006;98:681- 90.

34Hughes KS, Schnaper LA, Berry D, et al. Lumpectomy plus tamoxifen with or without irradiation in women 70 years of age or older with early breast cancer. N Engl J Med 2004;351:971-7
35 Sawaki M, Yamada A, Kumamaru H, et al. Clinicopathological characteristics, practical treatments, prognosis, and clinical issues of older breast cancer patients in Japan. Breast Cancer 2021;28:1-8.
36 Stueber TN, Diessner J, Bartmann C, et al. Effect of adjuvant radiotherapy in elderly patients with breast cancer. PLoS One 2020;15:e0229518.
37 Recht A, Comen EA, Fine RE, et al. Postmastectomy radiotherapy: An American Society of Clinical Oncology, American Society for Radiation Oncology, and Society of Surgical Oncology Focused Guideline Update. Ann Surg Oncol 2017;24:38-51

38 WILDIERS H, KUNKLER I, BIGANZOLI L, FRACHEBOUD J, VLASTOS G, BERNARD-MARTY C, et al. Management of breast cancer in elderly individuals: recommendations of the International Society of Geriatric Oncology. Lancet Oncol 2007;8:1101-15

39 BOUCHARDI C, RAPITI E, BLAGOJEVIC S, VLASTOS AT, VLASTOS G. Older female cancer patients: importance, causes and consequences of undertreatment. J Clin Oncol 2007;25:1858-69

40 BOUCHARDY C, RAPITI E, FIORETTA G, LAISSUE P, NEYROUD-CASPAR I, SCHAFFER P, et al. Undertreatment strongly decreases prognosis of breast cancer in elderly women J Clin Oncol 2003;21:3580-7.

41CLARKE M, COLLINS R, DARBY S, DAVIES S, ELPHINSTONE P, EVANS E, et al Effects of radiotherapy and of differences in the extent of surgery for early breast cancer on local recurrence and 15-year survival: an overview of the randomised trials. Lancet 2005;366:2087-10

42Yamada A, Kumamaru H, Shimizu C, et al. Systemic therapy and prognosis of older patients with stage II/III breast cancer: a large-scale analysis of the Japanese Breast Cancer Registry. Eur J Cancer 2021;154:157-66

43Muss HB, Berry DA, Cirrincione CT, et al. Adjuvant chemotherapy in older women with early-stage breast cancer. N Engl J Med 2009;360:2055-65.

44. Muss HB, Polley MC, Berry DA, et al. Randomized trial of standard adjuvant chemotherapy regimens versus capecitabine in older women with early breast cancer: 10-year update of the CALGB 49907 Trial. J Clin Oncol 2019;37:2338-48.

45 Jones S, Holmes FA, O'Shaughnessy J, et al. Docetaxel with cyclophosphamide is associated with an overall survival benefit compared with doxorubicin and cyclophosphamide: 7-year follow-up of US Oncology Research Trial 9735. J Clin Oncol 2009;27:1177-83.

46. Takabatake D, Taira N, Hara F, et al. Feasibility study of docetaxel with cyclophosphamide as adjuvant chemotherapy for Japanese breast cancer patients. Jpn J Clin Oncol 2009;39:478- 83.

47Crozier JA, Pezzi TA, Hodge C, et al. Addition of chemotherapy to local therapy in women aged 70 years or older with triple-negative breast cancer: a propensity-matched analysis. Lancet Oncol 2020;21:1611-9

48Paik S, Shak S, Tang G, et al. A multigene assay to predict recurrence of tamoxifen-treated, node-negative breast cancer. N Engl J Med 2004;351:2817-26

49Sparano JA, Gray RJ, Makower DF, et al. Prospective validation of a 21- gene expression assay in breast cancer. N Engl J Med 2015;373:2005-14
50Kalinsky K, Barlow WE, Gralow JR, et al. 21-gene assay to inform chemotherapy benefit in node-positive breast cancer. N Engl J Med 2021;385:2336-47
51Kizy S, Altman AM, Marmor S, et al. 21-gene recurrence score testing in the older population with estrogen receptor-positive breast cancer. J Geriatr Oncol 2019;10:322-9
52Romond EH, Perez EA, Bryant J, et al. Trastuzumab plus adjuvant chemotherapy for operable HER2-positive breast cancer. N Engl J Med 2005;353:1673-84.
53. Slamon D, Eiermann W, Robert N, et al. Adjuvant trastuzumab in HER2- positive breast cancer. N Engl J Med 2011;365:1273-83
54von Minckwitz G, Procter M, de Azambuja E, et al. Adjuvant pertuzumab and trastuzumab in early HER2-positive breast cancer. N Engl J Med 2017;377:122-31
55von Minckwitz G, Huang CS, Mano MS, et al. Trastuzumab emtansine for residual invasive HER2-positive breast cancer. N Engl J Med 2019;380:617-28.
56Sawaki M, Taira N, Uemura Y, et al. Randomized controlled trial of trastuzumab with or without chemotherapy for HER2-positive early breast cancer in older patients. J Clin Oncol 2020;38:3743-52.
57von Minckwitz G, Huang CS, Mano MS, et al. Trastuzumab emtansine for residual invasive HER2-positive breast cancer. N Engl J Med 2019;380:617-28

58Hortobagyi GN. Treatment of breast cancer. N Engl J Med 1998;339:974-84.

59Freedman RA, Tolaney SM. Efficacy and safety in older patient subsets in studies of endocrine monotherapy versus combination therapy in patients with HR+/HER2- advanced breast cancer: a review. Breast Cancer Res Treat 2018;167:607-14

60Park JH, Choi IS, Kim KH, et al. Treatment patterns and outcomes in elderly patients with metastatic breast cancer: a multicenter retrospective study. J Breast Cancer 2017;20:368-77 61 Cortes J, Cescon DW, Rugo HS, et al. Pembrolizumab plus chemotherapy versus placebo plus chemotherapy for previously untreated locally recurrent inoperable or metastatic triple- negative breast cancer (KEYNOTE-355): a randomised, placebo-controlled, double-blind, phase 3 clinical trial. Lancet 2020;396:1817-28

62Schmid P, Adams S, Rugo HS, et al. Atezolizumab and nab-paclitaxel in advanced triple- negative breast cancer. N Engl J Med 2018;379: 2108-21

63Nosaki K, Saka H, Hosomi Y, et al. Safety and efficacy of pembrolizumab monotherapy in elderly patients with PD-L1-positive advanced nonsmall-cell lung cancer: pooled analysis from the KEYNOTE-010, KEYNOTE-024, and KEYNOTE-042 studies. Lung Cancer 2019;135: 188-95

64Inoue K, Nakagami K, Mizutani M, et al. Randomized phase III trial of trastuzumab monotherapy followed by trastuzumab plus docetaxel versus trastuzumab plus docetaxel as first-line therapy in patients with HER2- positive metastatic breast cancer: the JO17360 Trial Group. Breast Cancer Res Treat 2010;119:127-36

65Kaufman PA, Brufsky AM, Mayer M, et al. Treatment patterns and clinical outcomes in elderly patients with HER2-positive metastatic breast cancer from the registHER observational study. Breast Cancer Res Treat 2012;135:875-83.

66 Sawaki M, Tamura K, Shimomura A, Taki Y, Nagashima F, Iwata H. Editors' Choice Practice management for elderly patients with breast cancer; findings from a survey by the Japan Breast Cancer Study Group. Nagoya J Med Sci 2018;80:217-26

67 Wildiers H, Tryfonidis K, Dal Lago L, et al. Pertuzumab and trastuzumab with or without metronomic chemotherapy for older patients with HER2- positive metastatic breast cancer (EORTC 75111-10114): an open-label, randomised, phase 2 trial from the Elderly Task Force/Breast Cancer Group. Lancet Oncol 2018;19:323-36.

68 Kirkhus L, Benth JS, Rostoft S, et al. Geriatric assessment is superior to oncologists' clinical judgment in identifying frailty. Br J Cancer 2017;117(4):470-7

69 Mohile SG, Dale W, Somerfield MR, et al. Practical assessment and management of vulnerabilities in older patients receiving chemotherapy: A.S.C.O. guideline for geriatric oncology. J Clin Oncol 2018;36(22):2326-47

70 Russo C, Giannotti C, Signori A, et al. Predictive values of two frailty screening tools in older patients with solid cancer: a comparison of SAOP2 and G8. Oncotarget 2018;9(80):35056–68.

71 Owusu C, Marggvucius S, Schluchter M, et al. Vulnerable elders survey and socioeconomic status predict functional decline and death among older women. with newly diagnosed nonmetastatic breast cancer. Cancer 2016;122(16): 2579-86

72Mueller CB, Ames F, Anderson GD. Breast cancer in 3,558 women: age as a significant determinant in the rate of dying and causes of death. Surgery 1978;83(2): 123-32.
73. Waldron RP, Donovan A, Drumm J, et al. Emergency presentation and mortality from colorectal cancer in the elderly. Br J Surg 1986;73(3):214-6.
74Fentiman IS. Are the elderly receiving appropriate cancer treatment? Ann Oncol 1996;7(7):657-8.
75Korc-Grodzicki B, Downey RJ, Shahrokni A, et al. Surgical considerations in older adults with cancer. J Clin Oncol 2014;32(24):2647-53.
76Martelli G, Miceli R, Daidone MG, et al. Axillary dissection versus no axillary dissection in elderly patients with breast cancer and no palpable axillary nodes: results after 15 years of follow-up. Ann Surg Oncol 2011;18(1):125-33.
77Giuliano AE, Hunt KK, Ballman KV, et al. Axillary dissection vs no axillary dissection in women with invasive breast cancer and sentinel node metastasis: a randomized clinical trial. JAMA 2011;305(6):569-75
78Chatzidaki P, et al. Perioperative complications of breast cancer surgery in oldern women (>/580 years). Ann Surg Oncol 2011;18(4):923-31
79Wyld L, Reed M, Collins K, et al. Impacts of omission of breast cancer surgery in older women with ER1 early breast cancer. In: 12th European Breast Cancer Conference. 2020.
80. Wyld L, Reed M, Collins K, et al. Cluster randomized trial to evaluate the clinical benefits of decision support interventions for older women with operable breast cancer. In: 12th European Breast

Cancer Conference. October 2-3, 2020; Virtual conference.
81Hughes KS, Schnaper LA, Bellon JR, et al. Lumpectomy plus tamoxifen with or without irradiation in women age 70 years or older with early breast cancer: long-term follow-up of CALGB 9343. J Clin Oncol 2013;31(19):2382-7
82Arriagada R, Mouriesse H, Sarrazin D, et al. Radiotherapy alone in breast cancer. I. Analysis of tumor parameters, tumor dose and local control: the experience of the Gustave-Roussy Institute and the Princess Margaret Hospital. Int J Radiat Oncol Biol Phys 1985;11(10):1751. 83 Martelli G, Miceli R, Daidone MG, Vetrella G, Cerrotta AM, Piromalli D, Agresti R (2011) Axillary dissection versus no axillary dissection in elderly patients with breast cancer and no palpable axillary nodes: results after 15 years of follow-up. Ann Surg Oncol 18:125-133
84Boughey JC, Haffty BG, Habermann EB, Hoskin TL, Goetz MP (2017) Has the time come to stop surgical staging of the axilla for all women age 70 years or older with hormone receptor- positive breast cancer? Ann Surg Oncol 24:614-617
85Chesney TR, Yin JX, Rajaee N, Tricco AC, Fyles AW, Acuna SA et al (2017) Tamoxifen with radiotherapy compared with Tamoxifen alone in elderly women with early-stage breast cancer treated with breast conserving surgery: a systematic review and meta-analysis. Radiother Oncol 123:1-9
86Vaidya JS, Wenz F, Bulsara M, Tobias JS, Joseph DJ, Saunders C, Brew-Graves C, Potyka I, Morris S, Vaidya HJ, Williams NR, Baum M (2016) An international randomised controlled trial to compare TARGeted Intraoperative radioTherapy (TARGIT) with conventional postoperative radiotherapy after breast-conserving surgery for women

with early-stage breast cancer (the TARGIT-A trial). Health Technol Assess 20:1-188

87 Saita Y, Ishijima M, Kaneko K. Atypical femoral fractures and bisphosphonate use: current evidence and clinical implications. Ther Adv Chronic Dis 2015; 6(4):185-93

88 He'bert-Croteau N, Brisson J, Latreille J, et al. Compliance with consensus recommendations for systemic therapy is associated with improved survival of women with node-negative breast cancer. J Clin Oncol 2004;22(18):3685-93.

89 Early Breast Cancer Trialists' Collaborative, G. Effects of chemotherapy and hormonal therapy for early breast cancer on recurrence and 15-year survival: an overview of the randomised trials. Lancet 2005;365(9472):1687–717.

90 Christman K, Muss HB, Case LD, et al. Chemotherapy of metastatic breast cancer in the elderly. The Piedmont Oncology Association experience [see comment]. JAMA 1992;268(1):57-62.

91 Muss HB, Berry DA, Cirrincione C, et al. Toxicity of older and younger patients treated with adjuvant chemotherapy for node-positive breast cancer: the Cancer and Leukemia Group B Experience. J Clin Oncol 2007;25(24):3699-704

92 Extermann M, Boler I, Reich RR, et al. Predicting the risk of chemotherapy toxicity in older patients: the chemotherapy risk assessment Scale for High-age patients (CRASH) score. Cancer 2012;118(13):3377-86.

93 Hurria A, Togawa K, Mohile SG, et al. Predicting chemotherapy toxicity in older adults with cancer: a prospective multicenter study. J Clin Oncol 2011;29(25): 3457-65.

94 Akhtar SS, Allan SG, Rodger A, et al. A 10-year experience of tamoxifen as primary treatment of breast cancer in 100 elderly and frail patients. Eur J Surg Oncol 1991;17(1):30-5
95 Jones S, Holmes FA, O'Shaughnessy J, et al. Docetaxel with Cyclophosphamide is associated with an overall survival Benefit Compared with Doxorubicin and Cyclophosphamide: 7-year follow-up of US oncology research trial 9735. J Clin Oncol 2009;27(8):1177-83
96 Kizy S, Altman AM, Marmor S, et al. 21-gene recurrence score testing in the older population with estrogen receptor-positive breast cancer. J Geriatr Oncol 2019; 10(2):322-9. 97 Shachar SS, Hurria A, Muss HB. Breast cancer in women older than 80 Years. J Oncol Pract 2016;12(2):123-32.
98 Perez EA, Romond EH, Suman VJ, et al. Trastuzumab plus adjuvant chemotherapy for human epidermal growth factor receptor 2-positive breast cancer: planned joint analysis of overall survival from NSABP B-31 and NCCTG N9831. J Clin Oncol 2014;32(33):3744-52. 99 Brunello A, Monfardini S, Crivellari D, et al. Multicenter analysis of activity and safety of trastuzumab plus chemotherapy in advanced breast cancer in elderly women (70 years). J Clin Oncol 2008;26(Suppl 15):1096.

100 Denegri A, Moccetti T, Moccetti M, et al. Cardiac toxicity of trastuzumab in elderly patients with breast cancer. J Geriatr Cardiol 2016;13(4):355-63.
101 Wishart GC, Bajdik CD, Dicks E, et al. PREDICT Plus: development and validation of a prognostic model for early breast cancer that includes HER2. Br J Cancer 2012;107(5):800-7. 102 Paridaens RJ, Gelber S, Cole BF, et al. Adjuvant! Online estimation of

chemotherapy efficacy when added to ovarian function suppression plus tamoxifen for premenopausal women with estrogen-receptor-positive breast cancer. Breast Cancer Res Treat 2010;123(1):303-10.

APPENDIX

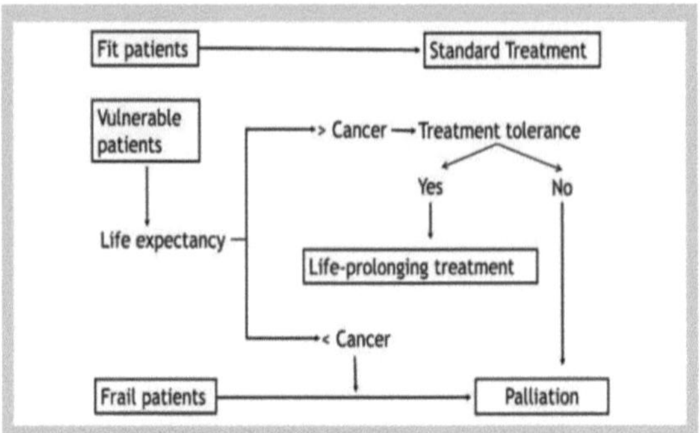

Figure 1. The treatment decision tree in older adults with cancer.

Figure 1:

MIX
Papier aus verantwortungsvollen Quellen
Paper from responsible sources
FSC® C105338

Printed by Books on Demand GmbH, Norderstedt / Germany